# SPECIAL FORCES

## Jim Brush

SEA-TO-SEA
Mankato Collingwood London

This edition first published in 2012 by
Sea-to-Sea Publications
Distributed by Black Rabbit Books
P.O. Box 3263, Mankato, Minnesota 56002

Printed in the United States of America, North Mankato, MN

9 8 7 6 5 4 3

Published by arrangement with the Watts Publishing Group Ltd, London.

A CIP catalog record for this book is available from the Library of Congress.

ISBN: 978-1-59771-294-1

Series editor: Adrian Cole
Art director: Jonathan Hair
Design: Simon Borrough
Picture research: Luped

Acknowledgments:
Photographer's Mate 1st Class Arlo Abrahamson/Photo Courtesy of U.S. Navy: 13b. Courtesy of American Technologies Network Corporation: 17br. Capt. Tommy Avilucea /AP/Press Association Images: 5t. John Berry/Image Works/Topfoto: 25c. © Boeing. All Rights Reserved: 17bl, 28. Patrick Chauvel/Sygma/Corbis: 22. Jordi Chias/Topfoto: 27. Chung Sung-Jun/Getty Images: 23c. Courtesy of Commonwealth of Australia: 8b. Corbis: 4cr. Mass Communication Specialist 1st Class Roger S. Duncan/Photo Courtesy of U.S. Navy: 14. Echoart/Dreamstime: 16cr. Romeo Gacad/AFP/Getty Images: 25t. Jim Gallagher/Crown Copyright/MOD: 29b. Sgt. David N. Gunn/Photo Courtesy of U.S. Army: 24cl. Chris Hondros/Getty Images: 24t. Mass Communication Specialist 2nd Class Erika N. Jones/Photo Courtesy of U.S. Navy: 10. MC2 Michelle Kapica/Photo Courtesy of U.S. Navy: 11t. SFC Andrew Kosterman/Photo Courtesy of U.S. Army: 18b. Mass Communication Specialist 2nd Class Dominique M. Lasco/Photo Courtesy of U.S. Navy: 13t. Staff Sgt. Jeremy T. Lock/Photo Courtesy of U.S. Navy: 20. Photographer's Mate 2nd Class Eric S. Logsdon/Photo Courtesy of U.S. Navy: 1, 11b, 26b. Greg E. Mathieson/Rex Features: 4t, 11c, 16cl. Paul Melcher/Rex Features: 17t. Pedro Jorge Henriques Monteiro/Shutterstock: front cover. Don Montgomery/Photo Courtesy of U.S. Navy: 29t. Scott Nelson/Getty Images: 9c. Robert Nickelsberg/Getty Images: 7b. Wessel Du Plooy/istockphoto: 18c. Reuters: 15b. Reuters/Corbis: 12. Nina Shannon/istockphoto: 16cra. Sipa Press/Rex Features: 19. Leif Skoogfors/Corbis: 8c. Sukree Sukplang/Reuters/Corbis: 15t. Courtesy of Thales Group: 23t. USAF/Getty Images: 7t, 21. Photo Courtesy of U.S. Army/Tech. Sgt. Jerry Morrison: 6. Photo Courtesy of U.S. Navy: 4cl. Won Dai-Yeon/AFP/Getty Images: 26c.

*Every attempt has been made to clear copyright.
Should there be any inadvertent omission please
apply to the publisher for rectification.*

November 2011
RD/9781597712910/002

**Author's note:**
**Some soldiers' faces in this book have been blurred to protect the identities of people still serving in the Special Forces.**

# Contents

Words highlighted in the text can be found in the glossary.

# Swords of Lightning

This is the "sword of lightning" badge of U.S. Army Green Berets.

**Special Forces are small groups of highly trained soldiers. They operate unseen and unheard behind enemy lines. Their enemies never know when or where they will strike.**

This is the badge of U.S. Navy SEALs. It shows an eagle holding a pistol, an anchor, and a trident.

## AF FACTS

Among the most famous Special Forces in the world are: the U.S. Delta Force, SEALs, Green Berets, and Rangers; the British SAS (Special Air Service); and the French Foreign Legion and Force d'Action Rapide (FAR).

U.S. Army Rangers using computers and communications equipment behind enemy lines.

Today, there are few large wars in the world, but there are many smaller **conflicts**. Special Forces teams strike hard and fast, like "swords of lightning." They can conduct an important mission without a country needing to send a large armed force.

*U.S.-trained Iraqi Special Forces carry out a hostage-rescue training mission in Baghdad.*

Special Forces teams are also sent in to deal with threats from terrorists, pirates, or well-armed criminal gangs. U.S. Special Forces also train local security forces to prevent attacks. Iraqi Special Forces have been trained to help prevent government officials from being killed or kidnapped.

## ACTION STATS

**The United States has the largest Special Forces—around 55,000 members in 2009. This number includes cooks, medics, mechanics, supply teams, and many other roles, as well as fighting soldiers.**

**"I accept the fact that as a Ranger my country expects me to move farther, faster, and fight harder than any other soldier."**
U.S. Army Ranger Code

# Who Are They?

Special Forces troops include men and women recruited from civilian life and from other armed services. They have years of training and experience. All candidates have to be physically and mentally tough, intelligent, skillful, and resourceful. They are the best at what they do.

*A Special Forces team practice attacking an enemy base.*

## AF FACTS

The British SAS is divided into units, each with their own special skills. There is a Boat Troop (diving), an Air Troop (parachuting), Mountain Troop (skiing), and Mobility Troop (long-range patrols).

Special Forces patrols are expected to fight against much larger forces. They must train harder than almost any other soldiers. Their motto is "train hard, fight easy." But they do a lot more than just fight. They need radio and language skills, first-aid training, plus **tactical** and survival skills in order to stay alive in tough environments.

This is the new MH-53J, equipped with an **infrared** system for night flights.

Special Forces teams usually work in secret. They are rarely seen by reporters or TV crews. Often the only way we know about their missions is after they have happened, or if any of the unit is killed during the mission.

## AF FACTS

MH-53 helicopters like this one are designed to land Special Forces troops in secret locations. They can fly close to the ground, even at night.

Some Special Forces units are rarely seen. This **sniper** is wearing a gillie suit, which has twigs and leaves added to help hide him.

"These guys are putting their lives on the line, taking on some very serious bad guys. The less anyone knows about the unit, the better."
Delta Force soldier

# Where Do They Fight?

Special Forces troops must be prepared to go anywhere in the world. Units are often sent on a mission at short notice in all sorts of terrain.

## AF FACTS

Navy Special Warfare forces were sent to Afghanistan in 2001. Within a year they had captured or destroyed more than 220 tons (200,000 kg) of enemy explosives and weapons.

U.S. Navy SEALs on patrol in Panama. They are part of a Special Boat Unit (see pages 26–27).

Australian Special Forces soldiers on patrol in Afghanistan.

Some Special Forces troops attack from the sea in fast boats, while others go into action by jumping out of planes. They operate in steamy jungles, scorching deserts, or icy mountains—some of the wildest terrains on Earth.

U.S. Special Forces soldiers use horses in Afghanistan to patrol areas where trucks cannot reach.

At all times, Special Forces troops must adapt to what is around them. In 2002, members of the U.S. Special Forces in Afghanistan rode around on horses. There were few roads, so it was the quickest way to move through the mountains.

## ACTION STATS

One problem for Special Forces teams is carrying enough food with them to eat. Fighting in the jungles of Borneo during the 1960s, Australian SAS troops lost up to 9 pounds (4 kg) in weight after a 12-day patrol.

# Tough Training

Basic training is very intense and can last many months. The first stage is fitness. Tough exercises include running, swimming, and long marches at night.

During training, **recruits** are given very few clues about what's in store for them the next day. A blackboard in the camp gives instructions, such as: "Red 3 report to vehicle at 0700 hours with full kit and a 50-pound (22-kg) pack." Recruits are tested on their own and in teams to see how well they cope.

## ACTION STATS

Recruits wishing to join the U.S. Delta Force must go on a 40-miles (65 km) march in two days, with little or no sleep. SAS candidates must climb a 3,000-ft. (880-m) peak three times in four hours.

*This recruit is just finishing a 5-mile (8-km) hike as part of a 15-week training program, before moving onto the next stage.*

**"On the first couple of speed marches, I only reached the end by being carried by the front men. I was exhausted."** British Parachute Regiment soldier

Recruits power their way through the surf on another training exercise.

Recruits attempt to swim 330 feet (100 m) with their hands tied. (Do NOT try this yourself!)

Training is not just about fitness. During SEALs "Hell Week" (above) recruits are taught the importance of listening to instructions

# AF FACTS

Only the best soldiers get to join Special Forces teams. Around 80% of candidates who apply or go through training are rejected.

# Skills Training

The second stage of training is six months of basic skills. Recruits learn how to move behind enemy lines, how to use different weapons, and how to work in teams.

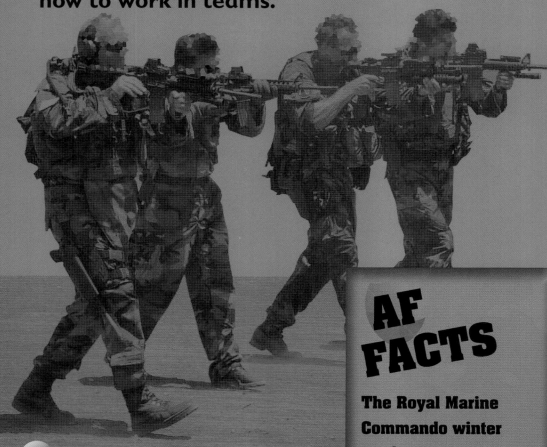

Members of the U.S. Army Special Forces show how to target the enemy while on the move.

Special Forces recruits can spend more than 1,000 hours of training just on the shooting range. They are also trained in basic first aid, tactics, using a radio, and handling enemy weapons.

## AF FACTS

The Royal Marine Commando winter warfare course includes a 186-mile (300-km) ski patrol. Recruits must enter "enemy" territory and gather information for later raids.

*Not only are recruits taught how to use radios to communicate, but also how to plot positions on maps and how to find their way there.*

The third and final stage of training is learning specialist skills, such as sniping, explosives, jungle warfare, or using a parachute. Some troops focus on learning how to drive in woods, deserts, or in the mountains.

*Skills training includes learning to use special vehicles, such as this Desert Patrol Vehicle (DPV).*

**"We aren't going to try to train you, we're going to try to kill you."** SAS training instructor

# Survival Skills

Special Forces soldiers must be able to survive behind enemy lines with little or no support. They are trained to find and build shelters and to locate food and water.

Shelters can range from a snow cave to a ditch covered in branches. Soldiers are also trained to light fires without matches. They must be able to find clean drinking water, and to live on fruit, nuts, roots, and wild animals. The U.S. Green Berets' nickname, "Snake-eaters," comes from their survival course.

*Making a warm fire in a frozen environment is just one of the key survival skills.*

During the 1950s, one SAS patrol in Malaya (now Malaysia) spent 103 days in a row in the jungle.

"My patrol commander was bitten on the arm by a scorpion and within a few hours, you would swear that someone had slipped a soccer ball under his skin. It was huge!"

Ian Conaghan, SAS trooper in the jungles of Borneo

A soldier catches a snake as part of a survival challenge.

Keeping weapons ready to use is another important survival skill.

# Basic Kit

**Most Special Forces units carry the basic equipment, such as a rifle, body armor, and a compass. They also carry electronic gadgets to help them fight at night or find their position.**

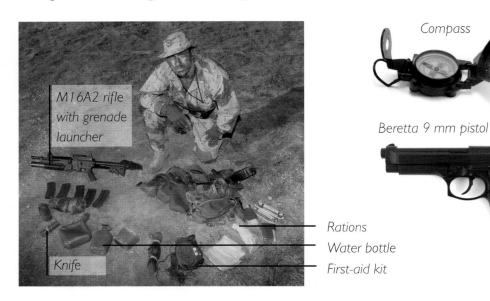

M16A2 rifle with grenade launcher

Knife

Rations

Water bottle

First-aid kit

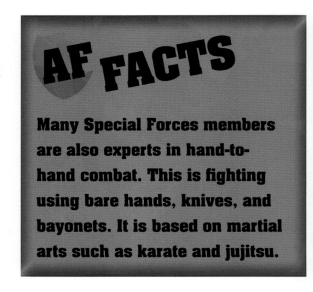

Compass

Beretta 9 mm pistol

Every team member carries a weapon for the specific mission, such as an M4 assault rifle, M24 sniper rifle, or Beretta 9 mm pistol. They also carry a combat knife, which can be used for hand-to-hand fighting (see right). As well as bringing water and rations (food), they also carry spare ammunition and radio batteries.

## AF FACTS

**Many Special Forces members are also experts in hand-to-hand combat. This is fighting using bare hands, knives, and bayonets. It is based on martial arts such as karate and jujitsu.**

# ACTION STATS

**A Global Positioning System, or GPS, shows a team's location, time, and speed. It uses signals from satellites in space. Wherever you are on Earth, it can show where you are standing within 165 feet (50 m)!**

Satellite smart phones, like this one, provide connections to military GPS networks.

Each team also has a GPS (see left). Radios and satellite phones are used to keep in touch with base. Night-vision goggles are used by teams to see when they operate in the dark. Other equipment, such as the CSEL (see below left) can help units when they are in trouble.

Night-vision goggles

The Combat Survivor Evader Locator (CSEL) uses a satellite network to link units on the ground with support and rescue teams back at base.

# Helicopter Drops

**A common way for Special Forces patrols to get behind enemy lines is to be dropped and picked up by small groups of helicopters.**

A helicopter's ability to land and take off in a small space helps Special Forces patrols land in secret locations. Helicopters can fly in all weathers, both night and day. They can also rescue patrols quickly if they get into trouble.

Five Puma helicopters take off on a mission, followed by their armed support—Apache attack helicopters.

## ACTION STATS

A CH-47 Chinook heavy-lifting helicopter can carry more than 50 Special Forces troops in one flight. The MH-47D is used by U.S. Special Forces and can be refueled in flight.

U.S. 1st Special Forces soldiers drop into the sea from a hovering MH-47D Chinook.

During Operation Telic in 2003, 97 Puma and Lynx helicopters landed 1,500 Royal Marine Commandos close to the Iraqi coastline. It took great skill to fly close to the ground at night through sand and dust storms.

**"Night dust landings were very difficult, despite having experienced soldiers."** Major General David H. Petreaus, commander of U.S. 101st Airborne Division

## ACTION STATS

**A Puma helicopter can operate day and night (when the crew wear night-vision goggles). It can carry 16 fully equipped troops or two tons of cargo. Six stretchers can be fitted for picking up wounded soldiers.**

*French commandos are dropped off by a Puma helicopter.*

# Jumping into Action

One of the main ways for Special Forces patrols to arrive in secret is to jump from planes at night. Using parachutes, they can glide quietly to the ground.

A squad carry out a HALO jump at 12,000 feet (3,600 m).

In High Altitude, Low Opening (HALO) jumps, teams jump out of the plane up to 33,000 feet (10,000 m) in the air. They make their body into a wing shape and "fly" toward their meeting point at speeds of 150 mph (240 km/h). At a set height, their parachute opens to slow their fall.

Members of the British Forces Special Tactics squad parachute to the ground following a HALO jump.

In High Altitude, High Opening (HAHO) jumps, the parachutes open right away. This allows the team to glide up to 30 miles (50 km) from their drop position. Special strips on top of the parachute allow members of the team to see one another.

## AF FACTS

Troops that parachute into the jungle wear a special harness. This allows them to parachute into the treetops, then lower themselves to the ground.

"The landing sites were usually chosen from a map... Whether you could land or not, you would never know until you got there."
NZ Special Forces trooper

# Mountain and Mobility

A soldier from Force d'Action Rapide lies ready for action in a snow-covered hole.

Mountain units—or mobility troops—specialize in mountain and arctic warfare. They are expert climbers and skiers. They can also handle a variety of vehicles, from a sled to a snowmobile.

Mountain units are trained to make parachute jumps into the snow. They also learn skills such as cutting trails through snow, rock climbing, and sliding down ropes, called rapelling. They also learn to "mountain walk," where they step over obstacles that could fall or cause avalanches.

*South Korean Special Warfare Forces ski during a snowy mountain exercise.*

Mobility troops are experts in using vehicles for hit-and-run attacks. They also have excellent desert survival skills. Each member needs to drink 2.5 gallons (9 liters) of water a day because of the hot sun. They often travel at night to stay hidden, and because it is cooler.

# Mission: Cave Fighting

In 2001, U.S. and British Special Forces teams were sent to attack terrorist training camps inside Afghanistan.

*Part of a destroyed cave suspected of being part of an underground training camp.*

*Members of U.S. 3rd Special Forces Group watch out for enemy forces in Afghanistan.*

It was a difficult mission but the teams had to work fast—winter was on its way. Many of the terrorist camps were inside large caves that were hard to spot from the air. Special Forces patrols on the ground found the caves, then called in planes for an airstrike.

## AF FACTS

In 2004, Willie Apiata, nicknamed "Mudguts," earned the Victoria Cross **while serving with the New Zealand SAS in Afghanistan. He** picked up his comrade and carried him 230 feet (70 m) across rocky ground, through enemy fire, to safety.

An airstrike on part of the Tora
Bora mountains in Afghanistan.

After the airstrike, Special
Forces patrols hid in the
rocks to capture the
survivors. On one occasion
they were surprised by a
large number of enemy
fighters who were hiding
in an underground bunker,
but they managed to fight
them off.

Special Forces work with local people, not only to
find terrorists, but also to help get food and equipment.

"SAS troops carried illuminated
bicycle flashers to mark themselves
if shot, so they could easily be
found in the smoke and confusion of

# Raiders from the Sea

Special Forces units such as the U.S. Navy SEALS attack from the sea. Patrols are carried to their target by submarines, fishing boats, landing craft, or canoes.

*Members of South Korea's navy commandos reach the seashore in their inflatable dinghy.*

*Special Boat Teams (SBTs) support SEAL units. This SBT is practicing high-speed, shallow-water control.*

## ACTION STATS

**The Craft-Riverine boat (below) used by U.S. SBTs can operate in very shallow water and has a top speed of 46 mph (75 km/h).**

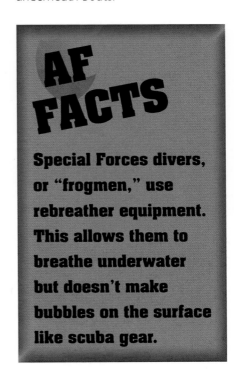

These Special Forces divers are practicing how to attach mines underneath boats.

Special Boat Teams are trained to find their way (navigate) at sea. They can use small boats, such as rigid raiders, zodiacs, and dinghies, as well as mini-submarines. These teams are used for scouting and for seizing oil rigs and ships captured by pirates or terrorists.

Some teams parachute into the sea with their canoes, then paddle to shore. Diving teams swim underwater and use mini-submarines to pull themselves along. During direct action, they sneak into enemy harbors and attach magnetic mines to enemy ships.

**"It can get hairy sometimes but I guess that's all part of the challenge, you have to be ready for it."**
SB1 N. Palmer, U.S. Navy

## AF FACTS

**Special Forces divers, or "frogmen," use rebreather equipment. This allows them to breathe underwater but doesn't make bubbles on the surface like scuba gear.**

# Who Dares Wins

**The challenges faced by the world's Special Forces are always changing. Lots of work goes into preparing for things that may happen in the future.**

The Boeing A-160 Hummingbird is a new unmanned rotorcraft—also called a UAV.

It's likely that Special Forces will remain the eyes and ears of the armed services. Nothing beats having a patrol on the ground to watch the enemy and report back to base. However, in the future, unmanned aerial vehicles (UAVs) may conduct some of their missions.

## AF FACTS

**The Hummingbird can fly more than 160 mph (258 km/h) and make decisions on its own. Eventually it could be used to support units on the ground.**

New equipment is being invented to help Special Forces teams in the future. This includes new boats and personal gear (see right).

**The SAS motto sums up the spirit of the Special Forces,** "Who Dares Wins."

# Fast Facts

● The U.S. Rangers were formed in 1754 under Major Robert Rogers. Traveling by foot and canoe across North America, they covered 400 miles (650 km) in 60 days without alerting their enemy, the Abernaki people.

● One of the most secretive Special Forces units is the British Special Boat Service, which has around 120 regular operators and a small number of reserves. They were part of a team that rescued *New York Times* journalist Stephen Farrell after he was taken hostage by Taliban fighters in September 2009.

● The British SAS has taken part in more conflicts than any other Special Forces unit, operating in Oman, Malaysia, Gambia, Falkland Islands, Colombia, Kuwait, Bosnia, Peru, Albania, Sierra Leone, Afghanistan, and Iraq.

● The South Korean counterterrorist unit, the 707th Special Missions Battalion, has a team of female operatives used for undercover operations.

● During the Cold War (1947–91), Russian Special Forces units, known as Spetznaz, had a fearful reputation. They were trained to assassinate senior NATO commanders and carry out attacks on missile bases and bridges.

# Glossary and Web Sites

**Bushcraft**—skills needed to survive in tough environments.

**Conflict**—an armed battle on a smaller scale than a full war.

**Frostbite**—frozen and blistered skin and tissue.

**Infrared**—light invisible to human eyes. Infrared vision equipment is used to see in the dark.

**Recruits**—people who have just signed up for military training.

**Resourceful**—thinking clearly and cleverly, even in tough conditions.

**Sniper**—a rifleman who hides himself to shoot at enemy soldiers, usually from a long distance.

**Tactical**—the use of skills involving moving and positioning to avoid or attack an enemy.

**Victoria Cross**—medal awarded to UK troops for extreme bravery.

**www.sealswcc.com**
Home of the U.S. Navy Seals who take their name from the elements in which they operate—Sea, Air, and Land. Find out about their history, download pictures, and watch training videos.

**www.howstuffworks.com/green-beret.htm**
and
**www.howstuffworks.comdelta-force.htm**
Everything you need to know about the Green Berets, the U.S. Army's Special Forces unit, and the even more secretive Delta Force.

**http://www.goarmy.com/ranger**
Find out what it takes to become a member of the U.S. 75th Ranger Regiment.

# Index